# AI FOR BREAKFAST

PUT AI TO WORK: REAL TASKS, DAILY HABITS, AND THE JUDGMENT TO USE IT WELL

AI AT THE TABLE
BOOK 2

THEO KERNEL

Published by Vector Verse Pty Ltd, Australia.

First edition, 2026.



This book explains real technology through an invented setting. The kitchen, its cooks, and every establishment in these pages are inventions; no real person, company, or product is named or depicted. How the book was made, and why nothing real is named, is described in "A Note on Names, and on How This Book Was Made."

This book provides general information about artificial intelligence. It is not professional advice, and nothing in it is legal, financial, or medical advice.

Paperback ISBN: 978-1-7648284-2-0

Vector Verse Pty Ltd · www.vectorverse.com.au

# A NOTE ON NAMES, AND ON HOW THIS BOOK WAS MADE

No company, product, model, lab, or famous name appears anywhere in these pages, and no dates. That is deliberate. A book that names this year's tools is out of date by next year's, and a book that leans on famous voices is borrowing authority rather than earning it. Nothing here is named, so nothing here can go stale and nothing here is owed. What is left is the only standard worth accepting: every claim is about how the kitchen works, put plainly enough that you can test it at your own counter.

Same kitchen, same rule, all the way down.

# A SMALL AI KIT TO START

## WHAT THIS BOOK IS

You've used one of these AI tools by now, the kind where you type a request and get back a draft, an answer, a paragraph. Maybe it impressed you; maybe it let you down; most likely both, on different days. What almost no one hands you is the part in between: how to cook with one of these things, day after day, on the real work in front of you. Knowing what these systems are and knowing how to work with one well are two different kinds of knowledge. This book is about the second.

This book is about cooking, in your kitchen: the everyday one where you do your own work. One word needs pinning down before we start. *Your* kitchen means your own working life, the desk and the tasks and the people you answer to. *The* kitchen, said on its own from here on, means the AI tool itself, the thing you cook with.

If you are arriving without *AI for Lunch* and *AI for Dinner*, here is all the ground you need: these systems cook plausible answers from a learned feel for how words usually go, built from mountains of examples; they are often useful and sometimes confidently wrong; and some of them can act, not just answer. You do not need the earlier books to use this one.

The stance this whole book takes has a name worth learning right away: the prep cook. The most productive way to work with these systems is not as a replacement for you and not as a gadget you refuse, but as a fast, tireless prep cook in your own kitchen. It does the chopping, the first drafts, the grunt work, at a speed you cannot match. You stay the chef: you direct it, judge its work, decide what is good enough, and own the result that goes out under your name.

Cooking well is mostly about working that relationship, and most of the failures come from getting it wrong in one of two directions, handing the prep cook your kitchen, or refusing to let it chop a single onion. That prep-cook stance, and the habit of always tasting before you serve, run through all three books of this series. This is the one where you put them to work.

So here is what I'm promising you by the last page. You will be able to put these systems to work in your job, productively, safely, without the confident mistakes reaching anything that matters, and without losing your own judgment and skills in the process. That last part matters as much as the rest. A tool that makes you faster while making you worse is a bad bargain.

The book runs like a tasting menu, each chapter a short course you can finish over breakfast. We start by stocking your kitchen and setting up, then move through the fundamentals of cooking with these tools. From there we work through the everyday dishes of professional life, code and pictures included. Then the stakes rise: proof you can no longer take at sight, the plates you send to others, the skills that rust when you stop using them, the jobs you keep for yourself, the prep cook let loose on whole jobs. We finish at your own kitchen table, with the recipes a team shares, and on becoming a better cook over time. Tie on your apron. Enough about what these tools are. Time to cook with one.

## STOCKING A SMALL KIT

A good home cook owns a small, well-chosen kit: a knife that fits the hand, a pan or two that do most of the work, a few specialized tools for the things she cooks often. Not one miracle gadget that supposedly does everything, and not a drawer crammed with fifty single-use tools never touched. The kit matches her cooking. Stocking your kitchen for working with AI is the same exercise, and getting it right saves you from both the gadget-for-everything trap and the cluttered-drawer one.

No single best kitchen exists, only the right one for the job. This chapter turns that into a practical starting move. Assemble a small kit of AI tools matched to your actual work, rather than defaulting to the one famous tool everyone names. For most people that kit is modest, a capable general assistant for the broad everyday work of words, perhaps one or two more specialized tools for things you do often, and an awareness of when a task wants a different kind of kitchen entirely. A few good tools you understand beat a pile you do not.

Choosing well means weighing a few plain factors against your situation. Capability: is this tool actually good at the kind of work you need, tested on your tasks rather than on a demo or a leaderboard? Fit: does it slot into how you already work, or fight it? Control and privacy: does it keep your data where sensitive work needs it to be? Cost: what does it take from you, in money and in the time you spend learning it? No single tool wins on all four.

The habit that settles a kit fastest is to test on your own tasks. Take a handful of the things you need done, the emails, the analysis, the drafts, none of it sensitive until a tool has cleared the privacy bar, and try them across a couple of tools. Your tasks are the only benchmark that measures what you need, and a short trial tells you more than any amount of marketing or any colleague's enthusiasm for a tool that does work nothing like yours.

I will give you the first lesson from stocking mine. Early on, I ran small kitchens locally, on my own machine, partly for the privacy and partly for the feel of it. They gave every impression of speed, answers streaming the moment I asked. Then I timed a real task against a big kitchen and the impression collapsed: what felt fast was much slower where it counted. The lesson outlived the tools. Test on your own dishes, with a clock, because the feel of a kitchen is a performance too.

Three habits produce a bad kit. The first defaults to one tool for everything, reaching for the single famous name even on tasks it is wrong for, because choosing is effort. The second chases the newest release, swapping tools for novelty rather than fit and never getting good at any of them. The third skips the test on your own work, adopting what others rave about for jobs unlike theirs. The antidote to all three is the same: think about your actual work, choose deliberately, and confirm by trying. A kit assembled by reflex serves you worse than one assembled on purpose.

A serious cook's kit reflects her cooking, chosen over time, each tool earning its place; an amateur's reflects the catalog and the latest fad.

# WHERE IT BELONGS, AND WHAT STAYS OUT

## FITTING THE PAN TO THE COOKING

Before a good cook turns on a burner, the work is already half-organized: ingredients prepped, tools laid out, the job fitted to the space and the equipment on hand. You do not rebuild your kitchen around a new pan; you find where the pan fits the cooking you already do. Putting AI to work in your job is the same move: not reorganizing your whole working life around the tool, but finding where it fits the work you already have, and setting it there.

This matters because the two most common mistakes with AI at work are to overdo it and to dodge it entirely. You know the colleague who has bolted it on everywhere, forcing the tool into tasks it only slows down, pushing it into every corner of the job because the hype said to, and who now has more friction than they started with. You may also know the one who never found a place for it, leaving genuine value untouched because fitting it in takes a little thought. The right move is in between. Find the specific places in your work where these tools help, and put them there.

Where do they help? The answer is short enough to carry in your head. The language-shaped parts of your work, the ones made of words, are where AI is worth reaching for when they also eat time and can be checked: the drafting you redo from scratch each time, the summarizing of long material, the reorganizing, the first passes, the explaining. The places it does not help are the ones that need exactness it lacks, judgment it cannot supply, or accountability only you can carry. The work is to map your own tasks onto that distinction, marking the spots where a fast prep cook would save you time, and the spots where you must cook yourself.

Start small and concrete. Pick one or two recurring tasks where you suspect AI would help: the weekly report you dread drafting, the inbox triage, the meeting notes. Integrate the tool there, properly, until it is a smooth part of how you do that work. A single well-fitted use that saves you time every week is worth more than a dozen half-attempted ones.

Two more errors of fit are worth naming. People often try to fit their work to the tool rather than the tool to their work, contorting how they do a job to accommodate AI, which is backward: the pan should serve the cooking. And they often judge the fit by novelty or guilt, using AI where they feel they should rather than where it helps, which produces busywork and calls it progress. The test for whether a use belongs is simple and honest: does it save you time or improve the result on a task that matters, checked against doing it your old way? If not, take it off the counter.

## WHAT NEVER GOES IN

One rule comes before all the dishes. A few ingredients never make it through the kitchen door. The cost is too high. At work, the equivalent is sharp and simple. Some things must never go into an AI tool you do not control, and knowing what they are, before you get into the habit of pasting, is the first line you draw.

Data hygiene rests on one plain fact about these tools. Whatever you type or upload goes to whoever runs the tool, and what they do with it, store it, study it, use it to train, is their policy, not your assumption. The things you might casually paste are often not yours to expose: confidential company information, client or customer data, anything covered by a confidentiality agreement or a regulation. Feeding that into a tool whose handling you have not cleared is not a small convenience; it can be a breach, with consequences beyond a bad plate.

Know your sensitive categories, then, and keep them out of uncleared kitchens. Before pasting, ask what this material is. Is it confidential? Is it someone else's private information? Is it regulated? If yes, it does not go into a tool you do not control or have not had cleared for that use. The inbox and the meeting notes are full of other people's words, so if you picked those tasks to start with, clearing the tool comes before building the habit. This is where a kitchen you run yourself stops being a preference. For sensitive material, a tool that keeps your data in-house may be the only acceptable one, and the convenience of the famous public tool is not worth the exposure. Watch the trickle, too: small pastes feel harmless one at a time, but many of them together can leak something you would never have sent all at once.

One word in that rule deserves unpacking, because it mostly gets used as a vibe: cleared. A tool is cleared when someone accountable, you for your own material, whoever owns the risk at work, has decided that this tool may hold this class of material, and would say so with their name attached. A tool everyone in the office happens to use is not cleared; it is merely common. And much of what flows through your work is other people's, a client's numbers, a colleague's draft, a customer's complaint, and their expectations travel with it. The clearance that matters is theirs as much as yours.

People take false comfort in the idea that the tool "forgets." The kitchen forgetting you between visits does not mean the company running it kept nothing. Your sensitive paste may sit in logs, in stored history, in training pipelines, regardless of whether the chatbot "remembers" you. At work, assume anything you put into a tool you do not control may be retained somewhere, and treat that assumption as the standing rule.

And it goes wrong in the ordinary pull of convenience against caution. The friction of doing a task yourself, or of using a slower cleared tool, tempts people to paste sensitive material into whatever is handy, because it is easier and the risk is invisible in the moment. That invisible-in-the-moment risk is the danger: nothing bad appears to happen when you paste the confidential document, right up until it does. The discipline has to be a standing habit, decided in advance, before the rush of a task decides for you.

A professional kitchen has hard rules about what never touches the food, decided in advance and followed without rethinking each time, because the pressure of service is no time to deliberate.

# BRIEF IT LIKE A COLLEAGUE

## WRITING THE ORDER TICKET

Now we cook. The skill that comes first, and improves every other, is writing a good order ticket. The principle is simple. In the kitchen of your own work, the difference between a vague ticket and a clear one is the difference between output you rewrite from scratch and output you can nearly use.

The kitchen cooks its entire answer around what you give it. The ticket steers the cooking; the rest comes from its palate, the feel for plausible text it built in training. So a vague request forces it to guess and a clear one sets up the plate. At work, the clear ticket means giving the tool what a competent colleague would need to do the task: what you want done, who it is for, what form and length it should take, any constraints or things to avoid, and, best of all, an example of what good looks like. The weak ticket, "write something about the delays," gets a generic guess. The strong ticket, "write a three-sentence status update for the team on the website launch: what shipped, what is left, and the one risk worth flagging, plain and calm," gets something you can almost send.

If you take one thing from this chapter, show an example

whenever you can. At work this is enormously powerful. So much of what you do has a voice and a shape that a paragraph of description cannot capture and one good example demonstrates instantly. Want a note that sounds like your team and not like a form letter? Paste one you were glad to send and ask for the same again. Want the analysis laid out the way your director reads it? Show it one that worked. It is much better at copying an example than at following a description. Handing it the pattern beats describing the pattern and hoping.

Cold: "draft an update on the migration for the team." What comes back is a competent nothing, three paragraphs of change-is-coming that would fit any project at any company. Now the same request with a strong ticket: the actual cutover date, the two systems affected, the one action people must take, the sentence your director used in Monday's meeting, and a note that half the audience is technical and half is not. Same tool, same twenty seconds of cooking. The second version comes back with the date in the first line, the action in bold, and one paragraph you will cut because it flatters the project. You are editing instead of starting over, and the whole gain came from what you put in the ticket, not from how you phrased the ask.

This skill compounds across your working day. Every task you hand the tool, the email, the summary, the draft, the analysis, goes better with a clear ticket. And you learn it by plain attention: notice when an output disappoints, re-read the ticket you gave, and you'll usually find the gap, a missing audience, an unstated format, no example, and you'll write the next one better. None of this is a trick of phrasing. It is knowing, before you ask, exactly what you want back.

A clear ticket gets the best the tool has; it cannot give the tool skills it lacks. A beautifully written request for exact arithmetic or current facts still needs the right tool, or your own verification. And a clearer ticket makes the output look more authoritative. Better prompting raises the polish, not necessarily the truth. Good ticket-writing makes the tool more useful. It doesn't make it trustworthy.

## STOCKING THE COUNTER

A clear ticket starts the dish, but the cook also needs the right things in front of her. Hand her the order and an empty counter and she improvises from memory; hand her the order plus the ingredients, the customer's note, the recipe to match, and she makes the real thing. What you put on the counter alongside your ticket often matters more than how you phrase the ticket, and learning to stock the counter well is where most of the quality in your AI work will come from.

Call this context engineering, brought to your daily tasks: gathering what the tool needs to see and putting it in front of the tool before it starts. The shift it asks is from treating the tool as a question box to treating it as a cook you are responsible for supplying. Most disappointing work output comes not from a bad prompt but from a bare one. You asked the tool to do something without giving it the materials the task needed. Stock the counter, the document to summarize, the customer's message to reply to, a strong past example to match, and the tool can do the job instead of a plausible generic version.

Take one of your own tasks. Asked cold to "draft a reply to this unhappy customer," the tool invents a generic apology that may contradict your policy. Given the customer's message, your real policy, and one past reply you were proud of, all stocked on the counter, then asked for the draft, it produces something specific and nearly usable. Same tool, same kitchen. What you put in front of it is the difference. The effective users of these systems stock the counter well.

This reframes much of what makes AI useful at work, and it runs beneath the everyday dishes we are about to cook. Summarizing well means stocking the document. Analyzing well means stocking the data. Replying well means stocking the thread and the relevant policy. In every case, the prompt sets the task and the stocked counter makes a good answer possible.

Stocking has its own limits, and they bite in three ways. The counter is finite, so stocking is curation, not dumping: pile on everything and you bury the important material and scatter the tool's attention. Stale or wrong materials produce confident wrong output with the authority of a cited source. And off-topic material gets worked in, because the tool assumes everything on the counter is there for a reason. So put the right, current, relevant materials in front of the tool, near your request, and leave the rest off the counter.

# TASTE BEFORE YOU SERVE

## THE DRAFT MINDSET

No cook sends the first attempt straight to the table. She plates it, looks, tastes, adjusts, and re-plates, and the dish that goes out is the second version. The most freeing habit you can bring to working with these tools is the same: the first plate is a draft. Not the answer, not the finished thing, but raw material to react to and refine. People who struggle with AI at work often expect one perfect output; people who get value from it expect a draft and improve it.

Iterating like this changes your relationship with the tool. Instead of writing one perfect request and accepting whatever comes back, you treat the exchange as a conversation: the tool produces a first attempt, you respond, "cut the second paragraph," "make it less formal," "you dropped the budget line," and it revises, and a few rounds of this usually beat one carefully composed request. The reason it works is that each pass still has the thread behind it. Every new version is shaped by the whole exchange, not just your latest note. At work, it means you stop trying to get it right in one shot and start steering toward right over a few quick rounds.

This habit is liberating because it takes the pressure off the first ticket and puts the value where it belongs, in the refinement. You don't need the perfect prompt; you need a good-enough start and the willingness to push back on what comes out. That is a lower bar to begin and a higher ceiling to reach. The draft mindset turns AI from a vending machine you hope dispenses the right thing into a collaborator you direct toward what you want.

It also keeps you properly in charge. When you treat the output as a draft you must shape, you stay engaged with the work, judging and directing rather than passively accepting. The draft is the prep cook's chopping; the plating, the final shape, the decision that it is good enough, remains yours. Iterating is how you use the tool's speed while keeping your hand on the result.

Steering goes wrong at both ends. Stop too soon and you're the cook who sends the first plate, settling for the mediocre generic version when two rounds would have gotten you something good. Stop too late and you are still muttering at the counter at round nine, chasing returns the draft stopped paying three rounds ago, when your own hand should have taken over to finish it. Knowing when to keep steering and when to plate it yourself comes with practice, like a cook's sense of when a dish is done.

## THE TASTING RULE

A serious kitchen has one rule it does not break: you taste the dish before it leaves your hands. At work it becomes a professional habit with your name attached, because here the plate you fail to taste does not just disappoint you, it goes out to a client, a colleague, or a boss, under your responsibility.

My own copy of this lesson cost me nothing, which is the cheapest way to buy it. I once asked one of these tools to make a photograph of me look professional, for a profile. The picture that came back was exactly what I ordered: better lit, better dressed, unmistakably professional. It just was not me. The face was close, confident, plausible, and wrong, and I caught it only because the subject was the one thing on earth I could verify at a glance. Everything else these tools hand me, I have to check on purpose.

At work, "serving" takes many forms, and the rule covers all of them: sending the email, submitting the report, acting on the analysis. Before any of that, you check. These systems produce confident output with no tell, fluent whether right or wrong. The only defense against an invisible mistake is to verify before you rely on it. At work, the invisible mistake is the wrong figure in the report you sent, the fabricated detail in the brief you filed, the confident claim to a client that turns out false, and the cost of those is professional.

The discipline scales with the stakes, and getting that proportion right is the craft of doing it without drowning. A throwaway brainstorm needs barely a glance. A draft you will rewrite anyway needs a light read. But some things need real verification: a number going into a report, a claim going out under your name, a decision with consequences. You check the figure, confirm the fact against a trusted source, and read the output as your own work, because once you serve it, it is your own work. You spend your checking where the cost of being wrong is high and relax where it is low.

What makes this discipline harder than it sounds runs through the chapters ahead. Call it the untasted-dish law: the dish you cannot taste is the one you most order in. You lean on the tool hardest for exactly the work you are least able to check, the unfamiliar subject, the document you did not read, the skill you never built. Its errors concentrate precisely where your defenses are thinnest.

The way to make this stick is to build the taste into the workflow as a fixed step. Decide, before you hand the tool a task that matters, what you will check and against what, and do not serve until you have.

Here is the whole discipline run once, at real size, so you can see what it costs and what it catches. Suppose you have handed the tool a thirty-page supplier agreement and asked for a summary before a renewal call. Five clean bullets come back. The check takes about four minutes and starts with a pen: circle every specific, each date, each figure, each named clause, because specifics are where wrong hides. First circle: the summary says the renewal falls on the fourteenth. Search the agreement, and the contract says the fourth, a stray digit that would have cost you ten days. Second: the summary reports the fee rising three percent. The clause says three percent or the index, whichever is higher, which is a different agreement altogether; the summary kept the half that sounded tidy. Third: the summary cites an addendum from last quarter. Search the packet; there is no addendum from last quarter. It is not hiding in small print, it does not exist, and a cited thing you cannot find is, for your purposes, not real. Fourth, and no amount of rereading the bullets would have caught it: the notice deadline, the one date the whole call turns on, appears nowhere. You find that not by inspecting what is on the plate but by asking what you needed the plate to hold.

Four minutes, four catches: a date off by ten days, a halved condition, a citation to nothing, and an omission. Notice what they share. Not one was visible on the surface of the summary, which read exactly as smooth wrong as it would have read right. Tasting, for this kind of work, is not looking harder at the dish; it is walking back to the pantry, to the source document, the original figure, the thing the output claims to be about. That is the whole procedure, four moves you can run on almost anything these tools hand you: circle the specifics, trace each one to its source, confirm that anything cited actually exists, and ask what is missing that you needed. How many of the four you run depends on the stakes, the same as everything else. On a throwaway draft, perhaps none. On anything with a date, a number, or a name in it, headed anywhere that matters, all four, and the four minutes is the price of the plate.

The tool works against you here. Its fluency and confidence make skipping the tasting feel safe: a polished, assured, well-formatted output feels checked. Professionals get burned by output that looked perfect and went out unchecked. The look of correctness did the job a real check should have done. The smoother the plate, the more you will want to wave it through, and the less reason you have to.

Verification is the price of the tool. Pay it every time you serve, or put the tool down.

The cook with the spoon is not suspicious of her own sauce; she is a professional, and untasted plates do not leave a professional's hands. Her name is on the plate, and at work your name is on what you send. That is one of the two laws this book keeps coming back to, and the one you will use every day.

# ORDER IN OR COOK IT YOURSELF

## THE PREP-COOK STANCE

Picture the line in the middle of a rush: heat coming off the griddle, tickets clipped in a row, one cook calling the orders and tasting every plate before it leaves, another at the board doing the chopping and the first hard sears. Naming that working relationship correctly is the difference between using these tools well and being used by them. The opening pages named the stance, the prep cook and the chef. This chapter is where it becomes a working division of labor. Hold that relationship, and the tool makes you faster and better. Lose that relationship, and it makes you worse.

The ticket, the counter, the draft, the tasting: each is one face of it, a way of directing a fast worker whose output you own. The prep-cook stance is the frame that holds the others together.

Concretely, the division of labor runs one way and not the other. The tool drafts the email; you decide what it should say and whether to send it. The tool summarizes the document; you confirm it kept what mattered. The tool generates options; you choose and refine. The speed is the tool's; the judgment and the accountability are yours, and that line does not move.

Holding this stance is what lets you capture the tool's enormous speed without surrendering the things that make your work yours. A prep cook makes a good chef faster; it does not make the chef unnecessary, and it does not get to plate on its own. When you keep that clear, AI multiplies you: the tedious volume lifted, your judgment freed for the parts that need it. The productive professionals neither hand everything over nor refuse the help. They run this relationship deliberately, all day.

People get this relationship wrong in two ways, and the damage is opposite each time. Treat the tool as a replacement rather than a prep cook and the failure writes itself: the judgment goes over with the chopping, the plates go out untasted, the decisions that were yours get made for you, and you end up with the confident mistake handed to a client and the quiet loss of the skills you stopped using. Refuse the prep cook entirely, out of distrust, and you lose hours of help on the safe, high-volume work the tool handles well. The first way carries the deeper cost, the deskilling risk: hand over so much that you can no longer judge or correct the tool's work, because you have lost the ability you were relying on it to replace.

The analogy holds, with one honest seam. Two costumes, one tool: the prep cook in your kitchen and the restaurant you order from are the same system, worn differently depending on how much of the job you hand across the counter. But the seam is this: a person on your line knows what the dish is for, and this one does not. It will keep chopping long after the dish has stopped making sense, at exactly the same speed, and it will never mention it. So the division of labor is right. What it leaves out is the supervision: nobody is watching the point of the work but you. Keep the chef's hands yours.

## THE ORDER-IN DECISION

Dinner comes three ways. Cook from scratch, when the cooking is the point or the result has to be exactly right. Order in, when speed matters more than control and good enough will do. Or do neither, when the honest answer is that this particular thing is not worth either effort. Telling them apart is the skill, and most people never develop it, defaulting to "order in for everything" or "never order in."

Handing a task to AI is the ordering-in option: one real choice among the three, with its own strengths and costs, to be matched to the job.

Order in, hand the job to the tool, when the task plays to its strengths and the downside is contained. It excels at first drafts, at brainstorming many options fast, at summarizing and reformatting. It excels, in short, at the language-shaped grunt work that swallows your afternoons. These are jobs where it is fast and useful, and where either the stakes are low or you can easily check the result. For that territory, refusing to order in is giving up hours you could have had.

Cook yourself, keep the tool out of it, or at least keep a firm hand on it, when the task needs what the tool lacks. The test is quick: could a confident wrong answer here cost you something you cannot get back, and would you catch it? Where the first answer is yes and the second is no, the kitchen is the wrong primary tool. And sometimes the cooking is the point: if a task exists partly to build your own understanding, ordering in defeats the purpose.

And sensitive territory stays out, on the data line already drawn: do not hand confidential or private material to a kitchen you do not control, regardless of how convenient ordering in would be. Some things never go through that window.

Most people get this wrong by habit, and let me guess which way you lean. You bought a book about using AI. You're not the holdout chopping every onion by hand on principle; you are the over-orderer, or you will be by spring, the first time a deadline meets a confident draft. So what follows is aimed at you, not at some imaginary purist. The skill is the match. Before you reach for the tool, ask: is this worth doing at all? If the honest answer is no, do neither and pocket the time. Is this language-shaped work where I can verify the result or the stakes are low? If yes, order in. Is this high-stakes, hard to verify, or something I need to think through myself? If yes, cook it yourself, or keep the kitchen under your own hand.

Ordering in is a clear, familiar choice with obvious trade-offs. The "is this a good job for AI" call is subtler, because the kitchen's confident fluency makes every job look like one it can handle, even the ones it cannot. So the judgment falls to you, and the chapters ahead are how you build it: knowing the verifiable from the unverifiable, and the task where speed wins from the task where being right is everything.

# IT DRAFTS FAST AND WRITES FLAT

## DRAFTING AND EDITING

The most common dish you will cook with this tool is words. Drafting an email, an outline, a memo, a post, a paragraph that will not start; or taking something you already wrote and cutting it, sharpening it, shifting its tone for a different reader. Writing is where most people meet AI at work, and it is, conveniently, one of the places where it helps most and where the trap is easiest to see, because you are the expert on what you meant to say. That last fact makes writing the right dish to learn first.

It works here because writing is cheap to check. When the tool drafts a paragraph, you can read it and judge it directly, because you know your audience, your point, and your voice better than it does. So the productive move is to use it to get past the blank page and through the tedious passes. Hand it a clear ticket, the reader, the purpose, the length, the tone, the points to hit, and let it produce a first draft you react to. Use it to cut a bloated paragraph in half, to make a stiff note warmer, to find the sharper word you cannot summon. It drafts and edits tirelessly, and on this kind of writing you can almost always taste the result yourself.

The weak spots are voice and fact. The tool writes in a flattened average: competent, smooth, and slightly anonymous, prone to hedging, padding, and the faint sheen that makes prose read as machine-made. Left alone, it will sand your point down to something blander and rounder than what you meant. Worse, it can insert claims you never made, a statistic, an assertion, a confident detail, showing up in the middle of an otherwise reasonable paragraph. That is what the trade calls hallucination, and it is worth being exact about it. This is not a glitch that strikes only where its produce was thin. The tool is always cooking for what sounds right rather than for what is so, which means a wrong claim can land inside a subject it knows well, plated with exactly the assurance it gives a correct one. So let it draft and revise while you keep two things firmly yours, the voice and the facts: use it for the friction, the blank page, the awkward transition, the third rewrite of a sentence, then read the result as an editor, not a reader, asking whether it sounds like you and whether every claim in it is true. The writers who use this tool well do not publish its drafts; they produce a draft fast and then make it their own.

A subtler cost accrues slowly. Publish its voice often enough and you stop noticing it isn't yours, and your own writing drifts toward that flattened average. Used without editing, the tool costs you more than a bland email today; it dulls the writer over months. Keeping the final voice yours is how you stay a writer while using a drafting machine.

Here is one of mine. We were promoting an event whose date had moved, and I asked for the copy for a poster. What came back was clean, confident, correctly formatted, and carried the old date: the thirtieth instead of the twenty-third. The tool was not inventing. It was working from an older version of the details, the way it always does when nobody puts the current ones in front of it. Nothing in the sentence looked wrong, because nothing in it was wrong except the one fact that mattered. Had it gone out, customers would have arrived a week late to a locked room, and the cost would have been paid in our reputation, not in my time.

It did not go out, and I would like to tell you I caught it. A colleague did, reading the poster against the thing it was supposed to be about, which is what I should have done. That is the part I have not forgotten. The check that saved me was not mine, and if he had been busy that afternoon the plate would have left under my name.

You need a short note telling the team a tool they rely on is being replaced next month. Ticket: what is changing, when, what they need to do about it, calm tone, under two hundred words, and say that most of the readers are not engineers. The draft arrives in seconds and it is usable: right shape, right length, wrong in three places only you can see. It buries the deadline in the middle paragraph; you move it to the first line. It promises the change will be painless, a claim you never made and cannot keep; you cut it. And its closing line is the flattened average of every migration email ever written; you replace it with the sentence you would actually say. The note is yours now: the tool supplied the volume, the editing supplied the truth of it. That is the whole shape of writing with it, run at full speed.

## SUMMARY AND SYNTHESIS

Some of the most useful work you will hand this tool is reduction: a long report you need the gist of, a thread of forty messages you need the decision from, six documents you need pulled into one view. This is making a stock, boiling a large volume down to its concentrated essence. Done well, it gives you back hours and lets you hold more in your head than you could read. Done carelessly, it hands you a confident essence of something you never checked, and you make a decision on a reduction that left out the one thing that mattered.

Two things make this work. First, put the source in front of it rather than relying on its memory, the stocking-the-counter habit from earlier: paste the document, the thread, the notes. Then it summarizes what is there instead of reconstructing from a vague recollection. Second, ask for the specific cut you need. "Summarize this" gets you a generic shrink; "pull out the decisions and who owns each," or "give me the argument and its main objection," or "list every figure mentioned," gets you a stock made for your dish. The tool compresses well when it has the material and a clear instruction about what to keep.

Summaries fail in what falls out silently. A summary's job is to drop things, and the tool drops based on a general sense of importance, not yours. It can silently omit the one detail you needed. It can shift emphasis, making a tentative point sound settled or burying the real conclusion. And if you did not give it the source, or gave it too little, it can summarize what such a document usually says rather than what this one does, inventing an essence. The seduction runs deepest here: you reach for a summary because you don't want to read the original, which is the same reason you can't check it.

Let the stakes set the rule. For low-stakes orientation, getting the rough shape of a long thread, a quick summary is fine and saves time. But for anything you will decide or act on, treat the summary as a guide to the original, not a replacement for it: let it tell you where to look, then read the parts that matter yourself. Never let a summary of something you never read be the basis of a consequential choice. The reduction points you to the passages it kept; it cannot point to what it dropped, so the higher the stakes, the more of the original you read yourself.

Synthesis across several sources carries the same warning, doubled. Pulling six documents into one view is valuable and, like any reduction, loses something. What it loses first is the disagreement: when the tool harmonizes sources that conflict, it hides the conflict. So add one line to the ticket, ask where the sources disagree, because the conflict was the point.

# RESEARCH IS WHERE IT INVENTS

## THE GIFT AND THE TRAP

The most dangerous thing this tool hands you in research is the source that is not what it seems. Sooner or later you will use it to learn something: to get oriented in an unfamiliar area, to have a concept explained, to find out how something works before a meeting. This is asking for ingredients you do not keep on hand. What arrives is either the real thing or a convincing imitation, depending on what you asked for, and getting that distinction right is the difference between a tool that makes you smarter and one that makes you confidently wrong in public.

It cooks well for the general and the established. Asking it to explain a concept, lay out how a process usually works, define a term, or orient you in a field you do not know plays to its strength, the well-rehearsed dishes it has effectively seen cooked thousands of times, and it's an excellent, patient tutor for this. Use it to build a quick map of unfamiliar territory: the main ideas, the vocabulary, the questions worth asking. As a way to reach the point where you can ask good questions, it is hard to beat.

Then you ask for something specific. The moment you need a

particular figure, a citation, or a recent event, you have stepped onto the ground where the tool invents most freely and most convincingly. A fabricated statistic looks identical to a real one. An invented citation has a plausible title, a plausible author, a plausible year, and does not exist. Sometimes it is worse than invented: a real author and title bent to support a claim they never made, which is harder to catch because the source itself checks out. The tool will not signal the difference, because to the tool there is no difference; it is producing the kind of thing that fits, and a real figure and a confident guess are the same kind of thing. This is not an occasional glitch. It is the predictable behavior of the machine at the edge of what it reliably knows.

The rule for research is fixed: use the tool for orientation and explanation, and a trusted source for any specific fact you will rely on. Let it build the map; verify the territory before you walk in. When it gives you a number, a citation, or a claim of fact that will end up in your work or your decision, that item is a lead to check, not a fact to use. Treat every specific the tool hands you as unconfirmed until you have seen it somewhere you trust. The orientation is the gift; the specifics are the trap.

You are weighing a vendor's claim and ask the tool what independent checks such a claim usually has to pass. It returns five, fluent and orderly: three you recognize, one certification with an official-sounding number, and one journal-sounding study with authors and a year. You treat all five as leads. The three familiar ones check out in minutes. The certification exists, but it covers a different class of product, a detail the tool's summary smoothed away. The study does not exist at all: the journal is real, the paper is not, and you would never have known from the citation, which looked exactly like its four neighbors. Ten minutes of checking, one invented source caught before it entered your memo, one real source rescued from a wrong reading. The tool did the sweep in seconds; the checking was still yours, and it was the part that made the work true.

Some versions of the tool reduce this risk by grounding their answers in real documents and showing you sources, so the kitchen looks something up before it answers. This helps, and it does not remove the need to verify. A grounded tool can still misread its source, or blend a real source with an invented detail. A citation shown is an invitation to check the citation, not proof it says what the tool claims. Even with sources on the screen, the specific you will rely on still gets your own look.

## WORKING WITH NUMBERS

Eventually you will point this tool at numbers: a column of figures to analyze, a calculation to run, a spreadsheet to make sense of. This corner of the kitchen holds the scale and the measuring spoons, where approximately right means wrong, and where one sharp distinction stops being abstract and starts costing money. The tool can be useful with data, but only if you understand which part of the numerical work it is good at and which part it must never be trusted to do alone.

What it does well is the structure of analysis. Reasoning about data is its ground: deciding what to compute, how to approach a messy dataset, which comparison would answer your question, how to lay out the steps. The tool is a capable thinking partner for all of that. Better still, it is strong at writing the thing that does the math: the spreadsheet formula, the small piece of code, the step-by-step procedure that, once you run it, computes the answer the same way every time. When the tool produces a formula and a calculator or program executes it, the arithmetic is done by something that actually does arithmetic, and the tool has done what it is good at, the setup, while leaving the computation to a machine that cannot misremember a sum.

It fails at doing the arithmetic itself, in its own voice, as part of generating text. The tool is a kitchen that learned the taste of correct-looking math, not a calculator. Ask it to add a column or compute a percentage inline and it will produce a number that looks plausible and may simply be wrong, with no internal check that the figures add up, because it is predicting a sensible-looking answer. It can be right often enough to lull you and wrong often enough to hurt you, and nothing in the tone tells you which you got.

Split the numerical work the way a careful kitchen splits tasks: let the tool reason about the data and set up the analysis, and have the computation done by a spreadsheet, a calculator, code you run. And because a wrong number looks exactly like a right one, a figure in a report or a client deck is the kind of plate you always taste: verify every number you will serve, tracing it back to a computation you trust.

"The tool gave me this figure" is not verification; it is the thing that needs verifying.

The deeper point concerns what kind of object you are dealing with. A calculator is a machine that is reliably right about arithmetic and knows nothing else; this tool is a machine that knows a great deal and is not reliably right about arithmetic. Confusing the two, treating the eloquent generalist as if it were the precise specialist, is the single most common way people get burned with data.

A great chef can design a recipe, refine it, and tell you exactly how the dish should taste; you still weigh the flour on a scale, because the chef's estimate and the scale's reading are different kinds of thing. Don't ask the tool to be the calculator; ask it to write the calculator.

# WRITING TO A PERSON, AND THINKING OUT LOUD

## WRITING TO PEOPLE

Most of the everyday writing you do isn't really about words; it's about a person. Some dishes are cooked for one person and no one else: an email to a touchy client, a note giving a colleague hard feedback, a reply that has to smooth a ruffled situation. They are seasoned for the regular who sends a plate back over too much pepper, and what matters is the right effect on a particular person in a particular relationship. An earlier chapter covered writing as craft; this is writing as communication, and the tool helps with it in bounded ways. Its help and its blind spot are closest together here.

It helps with the general parts of communication. Producing a serviceable first draft of a routine message saves time, and the tool is good at shifting register on demand: make it warmer, make it shorter, make it firmer without being rude. It earns its place on the message you have been avoiding, the awkward ask or the touchy reply, where its draft gives you something to react to. And it can offer a quick read on how a message might land. For the high-volume, routine mail that fills a workday, it is a capable assistant.

What the tool cannot see is the relationship itself, and that is where it fails you. It does not know the history between you and this person, the unspoken context, the thing you agreed last week, the sore subject to avoid, the particular way this colleague reads bluntness. So it writes the generically appropriate message, which can be subtly wrong for this specific person and moment: too formal with someone you are close to, oblivious to a dynamic you never told it about. The draft can read as exactly what it is: a competent stranger's guess at your relationship. A recipient who knows you will feel the difference without being able to name it.

This matters most where the stakes are highest. For the hard human messages, conflict, bad news, an apology, anything where care is the content, the generic version does not just underperform; it can do harm. The recipient needed to feel that you weighed your words, and a fluent default says the opposite. These are also the messages where the human point is that you wrote them: the effort and judgment are the message, and outsourcing them removes the thing that mattered. It can help you find words; it cannot carry the relationship, and it cannot be accountable for how the message lands.

It fails quietly, as the temptation to let the easy draft go out as-is. The convenience is real and the result is usually fine for routine notes, and that trains the reflex to send the tool's version, and that reflex then fires on the sensitive message too, where fine is not the bar. Scale your involvement to how much the message is about the relationship: use it freely for the routine and the purely practical, and take the pen back, using its draft as raw material at most, for anything where a specific person's feelings, trust, or your standing with them is the real cargo. The test is short: if what the message is worth depends on a person having written it, the tool can help you think but not speak.

## TALKING IT OVER

One use of the tool produces nothing you keep at all: thinking out loud. A cook works through a problem by talking it over, what if we tried it this way, what would go wrong, what are we missing, because saying it aloud and hearing it back sharpens the thinking. The tool partners tirelessly at this, available at any hour to generate options, poke holes, and play the other side, and used well it makes you think better. Used badly, it thinks for you, the opposite of the point.

Where it earns its keep is breadth and provocation. Ask it for fifteen approaches to a problem and you get a fast, wide spread, including a few angles you had not considered. Ask it to argue against your plan and it will surface objections worth answering. Ask it to help you reason through a tangle and the act of laying the problem out, and getting a structured reflection back, often clarifies your own view. Because you are using its output to spark and test your own rather than serving it, this stays low-stakes and forgiving, about the safest and most creative way to get value from the tool.

Underneath the breadth, the tool pulls toward the conventional. Its instinct is for the likely, and the likely is how things are usually said and done. Its ideas cluster around the obvious. Its "creativity" is frequently the predictable option in confident, polished phrasing that can feel more original than it is. If you mistake fluency for insight, you'll accept a well-worded ordinary idea as a good one, and you'll miss the unconventional answer that was the reason to think hard. The tool widens the field toward the middle, and it rarely reaches the edges, which is where the best ideas usually come from.

A second hazard is anchoring. Once the tool frames the problem a certain way and hands you its first ideas, your own thinking tends to snap to that frame, and you explore variations on its suggestions instead of stepping back to your own. This narrows exactly when you wanted to widen.

Both hazards come back to one confusion: the thinking partner mistaken for a thinking replacement. The value is that it helps you think; the failure is letting it think instead of you, taking its first plausible answer as the conclusion rather than as one input to your own. Worse, the habit erodes the very capacity you were trying to extend, and you cannot tell its good ideas from its average ones without your own judgment sharp. So use it to widen and to challenge, then do the judging yourself: treat its ideas as raw options to be weighed, push past its conventional first pass toward the angle it missed, and notice when you have anchored on its frame instead of your own. The good ideas are yours to find; the tool just makes the field bigger and argues with you while you look.

You have to tell a colleague their project is being folded into yours, and every draft in your head sounds like a victory lap. So you do not ask the tool to write anything. You describe the situation in three sentences and ask what the other person is likely to hear in each of three openings you are considering. It maps the readings quickly: the first sounds like blame, the second like pity, the third states the decision and leaves their standing intact. It also asks, flatly, what outcome you want the conversation to produce, a question you had not answered for yourself. You pick the third opening, write the message in your own words, and send nothing the tool wrote. Not one borrowed sentence, and the relationship walks away intact. The pen never left your hand; the thinking got wider anyway.

Bouncing ideas off a sharp colleague makes you smarter; handing the colleague the decision and signing whatever comes back makes you redundant. And this particular colleague's first instinct is usually the conventional one.

# LEARNING BY COOKING, AND THE SMALL STUFF

## THE TUTOR AT YOUR ELBOW

In the slow hour before service, an experienced cook lets a newer one work at her elbow: try the cut, taste the sauce as it builds, get a correction the moment it is needed with no one else watching. You can learn from these tools in that same way. Not look something up, which the research chapter covered, but build understanding or a skill over time: have a concept explained at your level, ask the follow-up you would be embarrassed to ask a person, try something and get feedback. As a tutor for the established and well-understood, the tool is remarkable, and it is reshaping how fast a motivated person can pick up a new area. It also carries a learning-specific danger that is sharper than it first appears.

As a patient, adaptive explainer it shines. It will explain a concept, then explain it differently when the first way did not land, answer the basic question without judgment, and meet you where you are stuck, all on demand and without tiring. It accelerates the broadly understood, the fundamentals of a field, the basics of a language, how a common framework works, collapsing the time from knowing nothing to knowing enough to practice. The patience alone is a gift, because much of learning is asking the same thing three times, and the tool never sighs.

Learning has a failure all its own, and it follows from the tool's confident inventing, the hallucination problem, with a twist. The tool can teach you something subtly wrong with complete confidence, and the untasted-dish law is at its sharpest here: not knowing the thing is why you came, so you are the person least able to catch it. When the tool misstates a fact, a method, or a fine point, an expert would notice and you can't. The error goes straight into your understanding and stays there, learned. Being taught wrong is worse than looking up a wrong fact, because it becomes part of how you think about the subject.

A second, quieter trap follows: the illusion of learning. Reading a clear explanation feels like understanding, and recognition isn't skill. Real learning takes the struggle, the practice, the trying and failing and adjusting that the tool can make it tempting to skip, because it is so easy to read another fluent explanation instead of doing the hard reps yourself. A subject you have only had explained to you is one you will not be able to use, and the deskilling concern applies in advance here: leaning on the tool to skip the effort means the skill never forms.

Both traps surface in practice, and the fix is the same for both: learn actively. Use the tool as a tutor to explain and to coach your practice, not as a substitute for the practice; verify what you learn against a trusted source as you go deeper, especially the foundations, since an error there is the costliest; and do the actual work yourself, the problems, the writing, the building, because that is where you build the understanding. The explanation begins the learning, and treating it as the end of the learning leaves you fluent-sounding and unable to perform.

Your team lives in spreadsheets and you have been bluffing your way past one formula family for a year. Ask the tool to explain it with one small worked example; the explanation is clear, patient, and feels like understanding. Now close the tool and build one yourself from a blank sheet. You stall at the second argument, and that stall is the actual lesson: the reading had installed recognition, not skill. Reopen the tool, name the exact step where you stopped, and have it coach that step only. Build two more from nothing. One honest failure, and the skill now exists in your hands instead of in your recent memory. Then, because the tutor can be confidently wrong, check its explanation of the tricky argument against the software's own documentation. It holds. It does not always.

The analogy is the apprentice and the master, and one difference must stay in plain view: this teacher explains patiently, corrects instantly, and can be confidently wrong, so the apprentice checks the lessons against the world in a way no apprentice ever had to before. But the deeper half is untouched: no apprentice ever became a cook by listening; she became one by cooking, badly at first, under a watchful eye.

## THE PREP COOK'S TERRITORY

Not every dish is a centerpiece. Much of what fills a working day is small: turning messy notes into a tidy list, drafting a routine reply, sorting information into order. Those minor tasks need doing, do not need your distinctive judgment, and take up whole afternoons. They form the prep cook's natural territory, where these tools sit at their safest and most reliably useful. After all the cautions, it is worth pausing on the easy win.

Low stakes and easy checking are the conditions where this shines. These are the tasks where errors are cheap and obvious: reformatting a block of text, generating a checklist or a template. The tool's fallibility barely costs you anything here. And the savings compound: offloading the dozens of tiny tasks that clutter a day frees attention for the work that needs you.

Even the small stuff carries three mild traps, and each has a short counter-move. The data line does not relax just because the task is trivial: pasting a sensitive document in to reformat it is still sending it out of the building, and a boring task is not a safe one if the content is confidential. The triviality itself invites the skipped check, and the wrong date in a "routine" calendar entry slips through because the task felt too minor to verify; give even small outputs a glance proportional to where they are going. And watch how far a small task can reach: one that felt trivial can carry consequences, like the routine message that happens to land on an important client, and the moment it does, it deserves the attention you give the things that are not small.

With those three held, lean on the tool freely for the low-stakes, high-volume work; this is what it is for, and refusing its help here, out of a caution meant for harder tasks, is the wasteful overcorrection. Most of the time the glance is a second, and the tool just gave you back ten minutes.

# CODE IT WROTE, PICTURES IT MADE

## CODE THAT LOOKS RIGHT

Consider a line cook who is astonishingly fast at the standard repertoire. Give him a familiar dish and it is plated almost before you finish ordering, clean and correct. He is so quick and so sure that you stop watching. That is the trap. Every so often he plates something flawless to the eye, with a flaw that surfaces only when the diner takes a bite. With food you would catch most of these. His actual specialty is writing code, and there the flaw can hide until the worst possible moment.

Coding assistants rank among the largest and most valuable uses of these systems. These kitchens learned from vast amounts of existing code, and they are strong at it: writing new code from a description, explaining what a piece of code does, finding and fixing problems, translating between programming languages. For people who write software, a good coding assistant is an accelerator, handling the routine and the boilerplate, the standard scaffolding every program needs, at a speed no human matches.

It works the way everything here works, by a learned feel, in this case a feel for what correct code tends to look like, drawn from its produce, the mountain of examples it learned from. And that phrasing should worry you by now. Code that looks correct differs from code that is correct. The kitchen produces plausible code, and plausible code can be subtly wrong, can carry a bug that runs fine until it doesn't, can be insecure in ways that stay invisible until someone exploits them. This is the hallucination problem again, now in code: confident, fluent, professional-looking output that may not do what it claims.

Looking right is the whole problem, because the error passes a glance. It runs strongest on common, well-documented patterns, where its produce was thick, and weaker and more error-prone on novel or niche problems, where it was thin, while sounding equally confident in both. It can introduce security flaws, especially dangerous because their whole nature is to be invisible. And it tends toward the common approach, not always the right one for your situation.

This makes the taste-before-serving rule literal and non-negotiable: you read the code, you run it, you test it, before you trust it. And it creates a sharp divide in who should use these tools and how. For someone who can read and test the output, a developer who treats the kitchen as a fast, fallible prep cook and verifies its work, a coding assistant is a large help. For someone who cannot evaluate what comes back, who pastes generated code into something important without the ability to check it, it is an untasted plate headed straight for the dining room. The one defense against confident wrong code is the ability to taste it. The tool amplifies the judgment of someone who has it and the danger for someone who does not.

The kitchen has no sense of when its code is off, and will plate the subtle bug with the same confidence as the correct function. The flaw is often invisible until the code runs in the wrong conditions, so testing that goes looking for those conditions is the real taste.

## IMAGE GENERATION

The same kitchen has another specialty that looks nothing like code. You type a sentence, a handful of words for a thing that does not exist, and a few seconds later a finished photograph of it is looking back at you: lit, composed, and yours. To most people this looks like pure magic, or like a thief who has copied someone's art, and it's neither. The same learned kitchen we have studied all along is working in a different medium.

It learned the same way, from examples: pictures with words attached, until it had a feel for how a description and an appearance go together. Hand it a new description and it makes a fresh image rather than fetching an existing one. But the motion at the stove is not the one you have been watching. Where a language kitchen lays down a sentence a piece at a time, this one starts with a full canvas of noise and lifts the noise away, pass after pass, until what is left fits your words, the way a shape emerges from fog as it thins.

Use it for concepts, drafts, and illustrations, and expect the failures specific to its medium; they are how you spot its work and avoid its traps. Image kitchens are notorious for certain tells: hands with the wrong number of fingers, text inside an image rendered as nonsense letters, details right at a glance and wrong on inspection. These happen because the kitchen has a feel for how things tend to look, not an understanding of how they actually are. It produces a hand that is plausible locally and impossible globally. It also struggles with precise requests for the exact arrangement of several specified things, because it is matching a feel.

Two heavier burns come with images. The first is that the kitchen learned from a vast body of pictures made by people who were never asked, which puts two questions in their most visible form: who consented to the pictures it learned from, and who owns what it makes. The second is that image generation is the doorway to synthetic media that can deceive, the deepfake problem we will face directly in the next chapter.

Where nothing turns on authenticity, order in freely. You can still catch some fakes by eye today, and knowing the giveaways helps, but treat that as a skill with an expiration date rather than a defense: the giveaways are fading, and the next chapter replaces eyework with the habit that keeps working. For anything where rights matter, do not assume a generated image is yours to use freely or clear of others' rights. And carry forward, into the next chapter, the unsettling fact that the same capability that makes a fun illustration can manufacture a convincing fake.

The analogy bends exactly where the fingers go wrong. A human painter understands that a hand has five fingers and a body occupies space; she paints from understanding. The kitchen paints from a feel for how images of hands tend to look, with no concept of a hand, which is why it produces six fingers with total confidence and no sense that anything is amiss. Keep the picture, a kitchen that paints from a description, and keep this with it: there is no understanding of the scene behind the paint, only a powerful feel for how such scenes usually appear.

# SEEING ISN'T PROOF ANYMORE

## PHOTOGRAPHS OF THINGS THAT NEVER HAPPENED

The kitchen can now plate a photograph of an event that never happened. A convincing image of a real-looking person in a place they never were, doing a thing they never did. Pair that with a learned voice and a moving image, and it can make that person appear to speak words they never said, in their own voice, on screen. This is the capability called deepfakes, or more broadly synthetic media, and it deserves a clear, calm chapter, because it changes something most of us have relied on our whole lives: the assumption that a photo, a recording, or a video is evidence that something real happened.

The mechanism is nothing new to you: the image generation we have already covered, the kitchen painting from a description, with the same trick producing speech in a learned voice, turned toward fabricating convincing fakes of specific real people and events. The same capability that makes a fun illustration or a helpful synthetic narration can manufacture a persuasive lie. A kitchen that can plate any scene from a description can plate scenes that never occurred.

A recording is no longer proof. For most of your life, a photo or a recording was as good as proof, because faking one convincingly was hard and rare. That is ending. We are moving into a world where a convincing image, audio clip, or video is no longer, by itself, proof that the thing depicted happened. Scammers exploit it for fraud, tricking people into believing a trusted person asked them to do something; propagandists use it to manufacture events that sway opinion; harassers use it to fabricate compromising media of real people. These are real harms, happening now, and they grow as the capability spreads.

Here is where it burns for you, not as a villain but as someone who has to decide what to believe and what to act on. The instinct that needs retiring is "seeing is believing." A sensational image or clip can no longer be trusted on its face, and the reflex that replaces it is not sharper eyes but a different way of checking.

## THE URGENT REQUEST

Start with the case that is already costing people money: the urgent message that sounds like someone you trust. A call in a familiar voice, a video note from a face you know, a message that needs you to act right now, send the payment, share the code, confirm the account. Anyone can copy a voice from a short sample of someone speaking, and the urgency is the tactic, because hurry is what stops you from checking. The defense is a habit, not a sharper ear. When a voice or a face you know asks you to do something that matters, confirm it through a separate channel you already trust: hang up and call back on the number you have always used, or reach them through an account you know is theirs. A simple family agreement helps too, a word or a question that a real person would know and a copied voice would not, kept for the moment when someone you love seems to be asking for help in a hurry.

## WHERE DID THIS COME FROM

A quieter harm costs people their judgment rather than their money: the striking image or clip that arrives already built to land. The more a piece of media wants you to feel something immediately, outrage, alarm, certainty about a person or an event, the more it has earned a pause. Before you believe it, and certainly before you pass it on, ask the plain question that has nothing to do with how real it looks: where did this come from. Does a source that would stand behind it actually stand behind it, or did it just appear, unattributed, perfectly tuned to make you angry. A real event tends to leave more than one trace, reported by people who can be held responsible for being wrong. A fabrication often arrives alone, with no origin you can follow.

What tells you now is not the polish but the provenance.

And one duty folds into this: every time someone forwards a fake without checking, they do the faker's work for free. Slowing down before you share is self-defense, and it's also how the problem stays small enough to live with.

Two cautions keep this from curdling. The first sharpens something the last chapter flagged: the giveaways are real but temporary. Do not build the habit on them. Today's fakes still slip sometimes, a hand with too many fingers, a blur where an ear should be, a voice that flattens oddly. Those giveaways are closing fast, and a defense built on catching them fails when the fake is good, which is the moment you needed it. Trust source and context, which don't get easier to fake as the technology improves. The second is the opposite error, and it is just as costly: deciding that because anything could be fake, nothing can be believed, which lets anyone wave away a real recording. The goal is not universal doubt but well-judged doubt, scaled to the stakes and to how hard the thing is working to provoke you.

None of this asks you to become an investigator. It asks for a small shift you can carry without effort: a voice, a face, or a photo is no longer proof by itself, so for anything that matters you confirm through a channel you trust and a source that can be held to account, and you give the loudest, most urgent media the least automatic trust. It costs you almost nothing on the ordinary day and protects you on the one day it counts.

# YOUR NAME TRAVELS WITH THE PLATE

## COOKING FOR THE NEXT STATION

So far the cooking has been yours alone: you make the dish, you taste it, you serve it, and if it comes back wrong it comes back to your station and nobody else's. But most work doesn't stop at you. The sauce your station makes goes onto someone else's protein; the analysis you finish feeds a colleague's decision; the draft you send becomes the basis of a document three people build on. The moment your output becomes someone else's input, the rules tighten, because now an error you fail to catch travels past you. Here using AI well stops being a personal habit and becomes a responsibility to the people downstream.

A handoff has a rule of its own, and it holds between humans as it held between you and the tool. Whenever one worker hands off to the next, the handoff is a place where a confident error can pass undetected and the receiver builds on it. The same thing happens on a team: you hand a colleague an AI-assisted figure, and they reasonably assume you checked it. The human handoff is as fallible as the machine one, and the polish of AI-assisted work makes it worse: output that looks finished invites more trust, not less.

You send a colleague a market summary you assembled with the tool, three pages, clean and confident, with one figure you meant to verify and did not. She trusts you, so she does not check it either. It goes into her slide, her slide goes into the regional pack, and six weeks later someone in another office builds a hiring plan on a number that was never real. Nobody was careless in a way anyone could point to. The figure simply passed four desks without ever meeting a source, because the first desk did not flag it and every desk after that read confidence as verification. A flagged figure would have stopped at desk two, cost one email, and been forgotten.

So the tasting bar rises with the reach of the plate. We said earlier to scale your checking to the stakes; feeding the team raises them. Before you pass along anything AI helped you make, the question is not only "is this good enough for me" but "is this solid enough for someone to build on without re-checking it," because that is what they will do. What you hand off, you vouch for, and you vouch for it whether the tool helped or not.

Taste harder on anything that feeds others, especially the specifics, the figures and the claims that a colleague will take as settled. And mark your uncertainty honestly when you pass work along: flagging that a number is still unverified, or that a section is a rough draft, costs you a sentence and saves the team from building on something unchecked. A clear "I haven't confirmed this figure yet" turns a hidden error into a visible open item.

The failure here turns on responsibility more than skill. When AI has a hand in the work, it becomes tempting to let accountability diffuse, to feel that the tool, not you, made the claim, so the error is somehow not quite yours. On a team this is corrosive, because everyone can lean on the same excuse and no one owns the result. "The tool generated it" is not a defense when your name is on the handoff. The work you pass along is yours, and a team where everyone remembers that catches errors a team leaning on the tool lets through.

On the line during service, every station trusts that what reaches it from the station before was made right, because there is no time to re-check upstream work mid-service. That trust lets the line move, and only each cook owning her handoff keeps it safe.

## WHEN TO DISCLOSE IT

An honest kitchen will tell you what is in the dish if you ask, and it must tell you when it matters, because a diner with an allergy has a right to know what they are eating. Using AI at work raises a quieter version of the same question: when, and how, do you tell people the tool was involved? No single rule covers it, since norms differ by workplace and by task, but one clear principle governs it, and getting it right protects both your integrity and the people relying on your work.

The principle comes down to honesty about what you're offering. Most of the time the tool's involvement is no more notable than which word processor you used: nobody needs a disclosure that AI helped draft an internal email, any more than they need to know you used spellcheck. But it matters when the recipient is relying on something the tool's involvement changes: when they are trusting your personal expertise, your original authorship, or your independent judgment, and the tool quietly supplied what they think you supplied. The line is not "did a tool touch this" but "does the recipient believe something about how this was made that is not true, and would it matter to them?"

That reframing settles most of the worry. Using the tool is not something to hide, any more than a chef hides the food processor; it belongs with the ordinary equipment. What you cannot do is misrepresent unchecked machine output as your own verified expertise. That is not a disclosure problem; it is a problem of competence and honesty, filed under the wrong heading. If a client is paying for your judgment and you hand them the tool's unexamined output dressed as your judgment, the breach is not that you used AI; it is that you did not apply the judgment they are paying for. Tasting the work and standing behind it is what makes it legitimately yours to present.

Follow your workplace's stated rules, which increasingly exist and override any personal preference. Where the rules are silent and you are unsure whether the tool's involvement would matter to the recipient, lean toward telling them, since the cost of unnecessary disclosure is small and the cost of a discovered concealment is large. And always, regardless of disclosure, do the work to make the output yours, because the disclosure question is far less important than the verification question behind it.

Concealment that misleads is the worst version: passing the tool's work off as expertise you did not exercise, which damages trust the moment it surfaces, and it usually surfaces. The remedy has its own quieter failure. Disclosure can become a substitute for diligence, the mistaken feeling that saying "AI helped with this" discharges your responsibility for it. It doesn't. Labeling a plate does not make it safe to eat; a disclosed but unchecked output is still unchecked, and the disclosure does not transfer the accountability back to the tool. The honest move is to verify first and disclose when it matters.

One more piece of honesty belongs in the dish: confidentiality runs the other way too. Just as you must be careful what you feed the tool, you owe the people whose information passes through your work the same care, and admitting "I put it through an AI" after the fact does not undo what you should never have done. Both halves of that are the same rule: tell people what is in the dish when it matters to them, and don't put things in the pot that were never yours to cook.

# THE SLOW RUST OF UNUSED SKILLS

## THE KNIFE YOU STOP PICKING UP

You have met this warning in pieces, the learning skipped here, the edge dulled there, a chapter's corner at a time. Here it is whole.

The quiet, long-term danger in these tools is not any single wrong answer, and it's easy to miss because it feels good all the way down. Picture the cook who, having a tireless prep cook for everything, gradually stops chopping, stops tasting, stops practicing, and discovers one day, when it matters, that the knife feels strange in the hand. Skills you stop using fade. If you let the tool do all of your writing, your analysis, your problem-solving, you slowly lose the ability to do those things yourself, and you may not notice until you need it and find it's gone.

The mechanism is not mysterious. Ability is built and kept by use, the way a cook's speed and judgment come from cooking, and outsourcing the work to a machine means you stop doing the reps that built the skill in the first place. For work you never want to do and were never good at, that may be a fine trade. But for the skills at the center of your craft, the writing if you are a writer, the analysis if you are an analyst, the reasoning that defines your value, trading away the practice is trading away the thing itself, slowly, in exchange for short-term ease. The cost is deferred and the convenience is immediate.

What makes this more than a general worry about getting rusty is the trap underneath it. Your ability to judge the tool's output depends on having the skill the tool is performing. A strong writer can see when a draft is weak; a capable analyst can smell when a number is off; the skill lets you taste the plate. So if you let the skill erode by handing everything to the tool, you lose the very thing you need to catch the tool's mistakes. That is the untasted-dish law, self-inflicted this time: you have made yourself least able to check the work you now hand over most. The deskilling does not just cost you the ability to work without the tool; it costs you the ability to use the tool safely.

The counter is not to refuse the tool. Use it to amplify skills you have rather than to avoid building ones you need, and deliberately keep your hand in on the abilities that matter most to your work. Do enough of the core work yourself to stay sharp, even when the tool could do it, because that practice keeps your judgment good enough to supervise. The chef who lets the prep cook do the chopping still picks up the knife often enough to know what good chopping feels like.

This shows up on two timeframes. For you, the erosion is gradual and invisible, you feel more productive every day even as the underlying skill thins, and you find out only when the tool is unavailable, or when damage surfaces from a mistake you could no longer detect. For those earlier in their careers, the danger is sharper still: people who reach for the tool before ever building the skill do not erode an ability; they never develop it, and they are left unable to judge the work they are, on paper, responsible for. Using the tool to learn faster is a gift; using it to skip the learning is a mortgage on your own competence.

A good kitchen uses every labor-saving device it can, and a good cook still keeps her knife skills sharp, because on the day a device turns out something wrong, the cook is the only thing standing between the mistake and the diner. So keep picking up the knife.

# THE JOBS YOU KEEP FOR YOURSELF

## TRUSTING THE CONFIDENT VOICE

The failure that undoes the rest is ours, not the tool's. We trust a confident machine more than we trust ourselves, and let a polished, assured output switch off the skepticism we'd bring to the very same claim from a person. A colleague says "the figure is around forty percent" and you ask how they know; the tool states it in clean prose and you paste it into the deck. Same claim, same uncertainty, completely different scrutiny, and the difference is not the evidence. It is that the machine's fluency disarmed you.

This tendency has a long history with every kind of automation: people defer to the automated answer, especially a confident and well-presented one, and stop applying their own judgment, sometimes even overriding what they can plainly see because the machine said otherwise. With these tools the pull is unusually strong, because their entire output is fluent, assured, and authoritative-sounding, the qualities that, in a person, we read as signs of competence. Confidence is not correctness, and this chapter takes up the human half of that problem: confident presentation buys trust it has not earned, from us, automatically, unless we actively refuse to extend it.

This is so corrosive because it attacks your defenses at the moment you need them. Every safeguard in this book, tasting before serving, verifying the specifics, checking the numbers, depends on you applying scrutiny, and this is the thing that switches the scrutiny off. It has a name, automation bias, and the bias in it is yours rather than the machine's: not a skew in what the tool produces, but a slackening in how you receive it. Worse, it strengthens as the tool gets better. A tool that is right nearly every time trains you, through long stretches of being correct, to stop checking, which is how the wrong answer, when it comes, reaches the plate unexamined.

The counter deliberately inverts the instinct, and it has to be deliberate because the instinct is automatic. Give the tool's output the scrutiny you would give a stranger's claim. Then give it more. A stranger can at least be embarrassed. The tool has never once been embarrassed. It will say the wrong thing in the same silk voice it used for the right thing, and it will rest easy, because there is no it. Make the check a fixed step that does not flex with how good the output looks, since how good it looks is the thing you cannot trust. The polish is not evidence. If anything, a particularly smooth and confident answer in a high-stakes spot deserves a second look, not a pass.

And here is the place oversight quietly dies. When a human is supposed to review the tool's work but has slid into trusting it, the review becomes a signature without a reading, a formality that adds the appearance of a safeguard while providing none. The team believes a person is reviewing the work; in practice a person is waving it through. That is more dangerous than no review, because it manufactures false confidence: everyone downstream believes the work was checked. A check you have stopped really performing is not a check.

The fanciest machine on a working line does not get to override the chef's own tongue; the chef tastes, however sophisticated the device that made the sauce, because the device cannot taste and she can. One thing more belongs to keeping the kitchen safe, and a confident cook forgets it: knowing which dishes not to attempt.

## DISHES NOT TO ATTEMPT

The rest of the damage does not come from using AI badly. It comes from using it at all on the one task that needed a human, handed over because the tool was there and it was easy. A cook turns down a dish when the night is wrong for it, and a good user of these tools knows which tasks not to hand them. This sounds backward, but it completes the lesson: knowing the edges means knowing where the tool does not belong, and choosing accordingly is what separates someone in command of it from someone reaching for it on reflex.

Start with what you already hold. The quick test from the order-in decision sorts most of this territory on its own: could a confident wrong answer cost you something you cannot get back, and would you catch it? To turn that test into a working list, spell out what it guards against, the losses you cannot get back, and the lines that sit alongside it. Keep the tool out when you cannot verify the output and the stakes are high, because then you are serving a plate you never tasted. Keep it out when the information cannot safely go in, the confidentiality line, where the right answer is simply not to feed it the task in that form. And keep it out when the task is the thinking you need to do yourself, the reasoning that builds your judgment.

The first case the test misses is the plain one the hype forgets: when doing it yourself is faster or better. The second is the licensed ground: your health, your legal position, your money. The tool holds no license and carries no accountability, and the gap between a plausible answer and a safe one is the gap a professional exists to close, so what it gives you is a starting point for a conversation with someone licensed, never the end of one. That line holds at home too, as a later chapter takes up; at work it hardens, because other people rely on what you decide.

The faster-yourself case is the most overlooked. Once the tool is at hand, every task starts to look like a task for it, the way every ingredient in the walk-in cooler starts to look like a job for the new machine. But setting up the request, reading the output, checking it, and fixing what it got wrong can easily take longer than just doing a small or familiar job yourself, and for work you know well, your own first pass is often better than the tool's draft plus your corrections. Reaching for the tool reflexively can cost you time while feeling like it saves time.

My own list has an entry that is not about stakes at all. Anything that has to be drawn: a process map, a diagram, a layout where the arrangement carries the meaning. I have tried this repeatedly, with several tools, and the pattern does not change. What comes back looks like a diagram, and then you start reading it. Boxes in the wrong order. A step invented to fill a gap. Two arrows where one belongs. Labels describing a process adjacent to mine. Fixing it takes longer than starting from a blank page, because first I have to work out what it thought I meant before I can correct it. So I draw those myself. Not on principle, and not because the stakes are high. Because the tool is worse at it than I am, and pretending otherwise costs me an afternoon.

The third turns less on cost than on what the work is for. Some tasks are valuable because a human did them: the judgment call that someone needs to own and stand behind, the difficult message where the point is that you took the care, the decision whose legitimacy comes from human responsibility. Handing those to a tool costs more than quality; it removes the person whose care was the entire value, and the output may look fine while missing the thing that mattered. Knowing which work is like this, and protecting it, is part of the judgment.

A manager asks you to draft the performance note for

someone whose work you have watched all year. The tool could produce something serviceable in twenty seconds, and you have all the material to feed it. You keep the writing anyway, not because the output would be bad but because the note's whole value is that a person who watched the work wrote down what he saw. So you use the tool for the shape, the standard sections, the order to cover them in, and you write every sentence about the person yourself. It costs ten minutes more. The difference is not visible on the page and it is the entire point.

The reflex itself is the problem: the habit of reaching for the tool without first asking whether you should, so that it creeps into tasks where it adds risk without adding value, or where the human point gets lost. And the cruelest version is reaching hardest for the work you can least check, because the task you cannot do yourself is the one you most want to hand off and the one you are least equipped to judge.

A chef earns respect not only for the dishes she sends out but for the ones she declines to attempt when the ingredients are not right or the risk is not worth it. Knowing the limits of the night is as much skill as cooking. Keep a short, honest list of the work you won't hand it. Underneath that list you need a map of where the tool's edges run.

## DRAWING YOUR OWN MAP OF THE EDGES

What you are building toward is an accurate, personal feel for where this tool is reliable and where it fails, on your actual work. An experienced cook has that for her own kitchen, in a way no manual could teach: which oven runs hot, which pan sticks, where the cold spot in the walk-in is. She learned it by cooking, by paying attention to where things went right and where they went wrong, until she had a precise map of her equipment's edges. The chapters before this one were groundwork for building that map.

This book gave you the general shape of the edges. These tools are strong on the well-rehearsed and the general, weaker on the specific, current, and precise; good at drafting and reducing, untrustworthy on facts, numbers, and citations; capable with the structure of a problem, unreliable on its exact details. That general map is worth holding. It only starts the job, though, because the edges shift by domain, by task, and by the particular tool. The map that protects you is the sharper one you draw from your own experience: where this tool consistently helps you, and where it has consistently burned you, on the specific work you do.

You build that map the way the cook built hers, by paying attention. Notice when the tool's output held up under checking and when it did not. Notice the kinds of tasks where you keep catching errors, and the kinds where it has earned a lighter touch. A lighter touch is still a touch: the streak that earns it is one you checked all the way through, and the streak you took on faith is the same one that trains you to stop checking. Over time that attention becomes calibration: a well-judged sense of how much to trust the tool for each kind of work. A novice trusts the tool the same everywhere and is therefore wrong about it half the time. An expert trusts it where it has earned trust and checks hard where it hasn't.

The habit is light but deliberate: keep a rough running sense, even informally, of your hits and misses with the tool, the tasks where it reliably saves you time and the tasks where it reliably costs you. Let that experience set how much you lean on it. And hold the map loosely. The edges move as the tools do.

Miscalibration has two faces you will meet in any office: the colleague who treats the tool as reliable where it is not, and lets confident errors past on the work she stopped checking, and the one who treats it as useless where it is strong, refusing help she could use and slowing everyone down for no reason. Both are failures of the same missing skill, an accurate sense of the edges, and both are common in people who have a fixed global opinion of the tool, good or bad, instead of a calibrated, task-by-task one. A further trap is generalizing across domains: the tool being excellent at your writing tells you little about whether it is reliable at your analysis, and assuming otherwise carries over a trust the tool never earned there.

The mark of a cook who knows her kitchen is not that she trusts her equipment or distrusts it, but that she knows what each piece will and will not do, and cooks accordingly.

# WHEN IT RUNS WITHOUT YOU

## WORKING WHILE YOU SLEEP

Everything so far has assumed you're standing right there while the tool works, each plate landing in front of you for a look before it goes anywhere. An agent is that tool handed a whole job instead of a single order, planning its own steps and carrying them out. The spot where you check the work before it leaves is what a restaurant calls the pass, the chef's last look on the way to the table. Much of an agent's promise is that it works when you aren't there: overnight, in the background, set off by something happening rather than by you asking. Think less of the prep cook at your elbow and more of a contractor with a key, let in before dawn to work an empty building and leave a note about what was done. An agent can run the same way, and that shift, from an agent you watch to one that runs on its own and reports back, changes both its usefulness and its risks.

These are background agents, sometimes called asynchronous. Rather than answering you in a live exchange, they run on a trigger or a schedule, do their work unattended, and surface a result when they're done. One watches for a certain kind of incoming request and drafts the response the moment it arrives; another runs each night to reconcile the day's records and flag the exceptions. Both take work off your morning, which is the gift, and both can run in the hours when no one is at the pass, which is the catch.

So for anything that runs unattended, lean hardest on the controls that need no human: hard limits, narrow powers, actions that are reversible or gated. Make the agent report its failures and surprises as loudly as its successes, and keep an inventory of what is running on your behalf. Whatever works while you sleep is what you are least likely to be watching when it goes wrong. A mistake made at three in the morning runs unchecked until someone reads the note.

## THE AUTONOMY DIAL

How closely you work with a capable cook covers a wide range. At one end, the sous-chef who works right under your eye: you are there, you see each move, you can say "not like that" before the plate goes out. At the other end, the caterer you hand the whole event to and walk away from, trusting them to run the night unwatched. One difference is worth naming before that range turns into a dial. A real caterer carries a license, a reputation, and someone to answer for it if the night goes wrong. An agent carries none of the three, so the far end of its range has to be earned with proof and never granted on the look of professionalism. With that said, both ends are valid. Which one you want depends on what is at stake, how much you trust the cook, and how much you can afford to have go wrong unattended. The choice is not whether to use a cook, but how much to hand over.

Agents live on this spectrum, and choosing your place on it is the central practical decision of using them. At one end is hands-on: the agent works under your eye, you see and approve its steps, and you catch problems early and cheaply. At the other end is hands-off, the autopilot: it runs the job on its own, and you see the result, not the steps. Autonomy is not a switch but a dial, and the dial sets the familiar trade: more autonomy is more power and more risk, together. Turn it toward autopilot and the usefulness and the exposure rise in the same motion; the mistakes happen out of sight, several steps deep, and you meet them at the end.

So let the cost of being wrong set the dial, not the thrill of automation. Stay hands-on when the stakes are high, the actions are irreversible, the agent is unproven, or you cannot easily verify the result; turn toward hands-off only for the low-stakes, reversible, well-proven, checkable work. And prefer an agent that raises its hand at the edge of its scope, the cook who knows which calls are above her station, while remembering that its sense of that edge is as fallible as everything else about it. Stay at the pass for the consequential steps regardless.

The two ends of the dial hold two opposite defaults, and the hype pushes hard toward one. Full autopilot for everything is reckless: it hands unwatched autonomy to jobs whose mistakes you cannot afford. Hands-on for everything is merely wasteful: the leash so short that the agent saves you nothing, and you have bought a chatbot with a longer receipt. Start any new agent toward the hands-on end, watch it earn trust, and turn the dial only as far as demonstrated reliability and acceptable stakes allow.

Hands-on first, hands-off earned.

## SCOPING THE JOB

The most common way agents disappoint in practice is not a failure of the machinery; what fails is the scoping. Hand a kitchen a clear catering order, sixty guests, set menu, dietary notes attached, delivery at seven, and a competent kitchen will nail it. Hand the same kitchen "make everyone happy somehow," and watch it flounder, not because it is a bad kitchen, but because the job is unscoped. The difference between an agent that delivers and one that disappoints is, more often than anything about the agent, the shape of the job you handed it.

A job suited to an agent has a clear goal with a definite finish line, steps it can sensibly be broken into, and a result you can verify. Reversible actions, or gates where they are not reversible, come with the package. The mirror image belongs nowhere near an unwatched agent: vague open-ended goals, long chains of steps, and anything needing judgment the tool does not have. So shape the task before you hand it over, give it the right tools and no more, and wall it off from anything live until it is proven. And do not pay your most expensive cook for bread service. The big, costly kitchen earns its price only on the steps that need it.

You want an agent to keep a supplier list current: check each supplier's published terms weekly, note what changed, flag anything that moved by more than a small margin. Clear goal, definite finish line, steps that break apart cleanly, everything reversible, results you can spot-check against the source in two minutes. That job is well shaped, and it will probably work. Now add six words: keep the list current and update our contracts to match. Same easy sentence, different animal. The finish line is gone, the actions are no longer reversible, and a wrong reading of a clause now rewrites an agreement. The first version is an errand. The second is a decision wearing an errand's clothes.

People hand agents jobs that are too vague, too long, or too high-stakes, then conclude that agents do not work. Agents today are good at well-scoped, error-tolerant, verifiable jobs and unreliable at sprawling autonomous ones, and which kind you hand them is usually yours to control.

The data line from the start of this book does not relax because the cook is now an agent; it tightens. An agent reaches your files, your messages, your tools, and if it runs on someone else's service, what it reads can leave the building with it. Decide what such an agent may touch with the same advance discipline you used for what never gets pasted, and treat a vendor's agent as a vendor: know where the data goes before the convenience decides for you.

## COUNTING THE WHOLE BILL

Two last pieces of arithmetic. If you're choosing between building your own agent and using one that exists, remember that the demo is the easy part and the bill comes later. Keeping an agent reliable as everything shifts beneath it is the never-ending remainder. The cost of building is rarely the building. It is the feeding of what you built, every day after.

And whatever you run, count the whole bill. The true cost of an agent folds in the setup, the fitting into your work, the mistakes it will sometimes make, and above all the watching: the agent that appears to save you an hour may eat forty minutes of your attention in review. The pass is not free; the human at it costs time, and that time is part of the price of the dish. The book has now asked for the same arithmetic at three sizes: the new fit that has to beat your old way, the per-task overhead that can make doing a small job yourself faster, and the running bill of an agent on the clock. One test covers all three: the honest total against the honest alternative. The goal was never to make the agent do the work. The goal is for the total value to beat the total cost, honestly counted, and some jobs fail that test while feeling modern.

# AI AT YOUR KITCHEN TABLE

## THE SMALLER, WARMER KITCHEN

This book has spent nearly all its pages at your desk, on the work that carries your name. But the same tool rides in your pocket on the weekend, into a smaller and warmer kitchen, the one with the radio on and nobody to impress. The everyday uses run just as you would guess, and they need no separate lesson. Two things about home do.

First, home is the best place to get good at this. The stakes are low, the feedback is immediate, and no one is watching. The person who has spent a few months asking the tool to explain things, draft things, and plan things in ordinary life arrives at the high-stakes work already fluent: already used to writing a clear request, already expecting a draft rather than an oracle, already tasting before trusting. Skill with these tools is a feel you build on the easy things and carry to the hard ones.

Draw one line clearly here, especially for students and for their parents, between using the tool to understand and using it to hand something in. Ask it to explain how a kind of problem works and you learn; ask it to do the problem for you and you have skipped the very learning the task existed to produce. The thinking you outsource while you are still learning is the thinking you never build. Pointed at teaching you, it is as patient a tutor as you will ever have. Pointed at doing the work in your place, it robs you of the skill the work was meant to grow. The same holds for any ability you are trying to keep sharp, your own writing, your own reasoning, your own memory for the things you mean to carry yourself.

Much of this now happens by voice and on the move, spoken rather than typed, in the car, at the stove, on a walk. The ease is real, and it changes none of the rules. A spoken answer is as fallible as a written one, and the easy, conversational feel of talking to the thing makes that easier to forget. The same check on anything that matters still applies.

Second, at home nobody has drawn the lines for you. At work, someone may have told you which tools your employer has cleared and which data are off limits. At home the only guard is you, and the things you're most tempted to type are often the most sensitive you have: your symptoms, your history, your finances, the details of your children. The same goes for the highest-stakes questions a life throws up, the health worry at midnight, the money decision, the legal bind: the tool can help you frame what to ask, and it has no license to lose and no duty of care to you, so on questions like these it is a place to start thinking and never the place to stop. Decide your home data line once, in advance, and keep the genuinely private, especially other people's and your children's, out of a service you have not cleared.

So the weekend isn't a lesser version of the work. It's where the habits are cheap to build, and the one place where every line is yours to draw.

# HOUSE RECIPES FOR THE WHOLE TEAM

## BRINGING THE TEAM ALONG

A tool delivers value only when people use it well, and people are the hard part, the part the technology most wants you to forget. A new technique changes a restaurant only when every cook on the line actually cooks it, not when the chef demonstrates it once to applause. If the line cooks don't adopt it, or adopt it badly, the technique is a story the chef tells, not a change in what leaves the kitchen. Standing the tool up is the easy half; getting a team to use it well is the half that decides the outcome.

You know the two ditches by now, and a team can be in both at once: the tool left unused out of fear, distrust, or plain inertia even as its cost accrues, and the tool over-trusted, its confident output waved through unverified, the familiar problem now multiplied across a group. Hand out access and hope, and you get a scatter of both rather than the calibrated middle.

That calibrated middle is the goal. It looks like people who understand what the tool is good at and where it fails, who trust it about as much as it deserves on each kind of task, and who follow shared practices for the things that must not be left to individual whim, the verification standards and the sensitive-data lines especially. Getting there takes the unglamorous work of training, clear norms, and time. A team learns well-judged trust the way an individual does, through guided experience with the tool's hits and misses.

A team of nine gets access on a Monday; by Friday two people are using the tool for everything, five have tried it twice, and two have decided it is a toy. Nobody is wrong exactly, and the output quality across the team now varies more than it did before the tool arrived. Six weeks later someone pastes a client list into a service nobody cleared, because no one ever said which service was cleared or that pasting client lists was the line. The fix was never a better tool. It was one page: here is what we use it for, here is what never goes in, here is what you check before it leaves your desk. It would have caught the only thing that actually went wrong.

## WRITING THE HOUSE RECIPES

A serious kitchen doesn't make every cook rediscover how to cook the signature dish. It writes the recipe down, shares it, and trains to it, so that every plate that leaves tastes the same no matter who is on the line. When a team uses AI, the same move turns scattered individual experiments into a shared capability, and skipping it leaves a group where everyone repeats the same mistakes in private. So the work of bringing the team along begins with writing the house recipes for cooking with these tools.

The opportunity is real and routinely missed. Across a team, people are independently working out the same things: which tasks the tool handles well, what a good instruction looks like for a recurring job, where the tool tends to go wrong, how much checking a given output needs. Left private, that knowledge stays in one person's notes and leaves when they do. Written down and shared, it compounds: a well-crafted instruction for a recurring task becomes a recipe the team reuses, an agreed list of what the tool is trusted for saves everyone the same trial and error, and a new colleague inherits the team's judgment instead of rebuilding it from scratch. The clear-ticket and stocked-counter habits from an earlier chapter are far more valuable as shared house recipes than as private ones.

But two house recipes have to belong to the team rather than to individual improvisation, and they are the safety ones. The first is what nobody feeds the tool, the confidentiality and sensitive-data lines from earlier, which must be an explicit, shared, written standard. The weakest individual judgment becomes the group's exposure: one careless paste, or one unverified output reaching a customer, creates a problem for everyone. The second is what must always be tasted, the verification expectations for work that carries real stakes, so that nobody ever has to say "I assumed someone else checked it." Personal habits are not enough for these two; they have to be team rules, stated out loud, because their failure is collective. Adoption work that builds enthusiasm without first settling those two rules raises the team's exposure as fast as it raises the team's usage.

Make the implicit explicit, and keep it light. Capture the instructions and approaches that work for your recurring tasks so they can be reused; agree as a team on which jobs you trust the tool for and which a person always checks; and write down the two safety lines, what stays out of the pot and what always gets tasted, as standards rather than suggestions. None of this needs to be elaborate. A short, living page that the team follows beats a thick policy nobody reads, the way a smudged index card that gets used beats a beautiful binder on a shelf.

This one costs you something on each side. Without shared recipes, a team gets inconsistent quality, repeated avoidable mistakes, and no common understanding of the lines, every cook improvising a slightly different dish. But standardizing too hard costs something too: write rigid, heavy rules that don't fit the work and people ignore them or work around them, which leaves a policy everyone violates and no practice underneath. And a team can standardize a bad habit as easily as a good one. A group that has quietly agreed to skip the spoon is more dangerous than one careless individual, because the bad practice now has the authority of the house behind it.

Most teams lose this to the deploy-and-hope reflex. Assuming people will naturally use a powerful, fallible tool well is the mistake; they won't, not because they are careless but because calibrated use is hard and has to be learned. Two early-experience traps make it worse: a few bad first encounters can sour a team into blanket distrust, and a few good ones can breed the complacency that drops the verification right before the costly miss. Treating adoption as a real effort, with training and patience, is what turns a deployed tool into a used one and a used one into a safely used one.

One last piece remains, quieter than adoption and just as real. The tools themselves will not hold still, which is why that short page has to stay living.

## KEEPING UP AS TOOLS CHANGE

These tools change faster than any kitchen gadget ever did, and the skill of keeping up without being jerked around is its own discipline. Kitchen gadgets come and go: every season brings a new device promising to change everything, and most end up in a drawer. The fundamentals of cooking stay as they were: heat, salt, time, taste. The cook who chases every new gadget masters none; the cook who refuses to look at anything new eventually falls behind a craft that does move.

Separate the fast layer from the slow one. The specific tools, their features, their interfaces, and the exact list of what they can and cannot do will keep changing, sometimes monthly, and trying to anchor your understanding to today's specifics guarantees you will be out of date soon. But underneath that fast-moving surface sits a slow layer that barely moves, and it is the one you have been building all along.

This gives you a stable way to judge anything new, instead of being pushed around by each announcement. When a new capability appears, do not ask whether it is impressive; ask the same fundamental questions you would ask of anything: what is it reliable at, where does it still fail, and what new verification does it require? A genuine advance answers those questions well and changes your calibrated map in a specific way. A repackaged claim does not survive the questions. Judging new tools by the fundamentals rather than the marketing is what keeps you current without making you credulous.

As the tools move, two edges of your map shift. The tool may become reliable at something it used to fail at. A task you correctly kept out of the kitchen last year may be worth handing over now, and refusing to revisit your map freezes you into outdated caution. Equally, a new capability may be claimed but not real, or real but narrower than advertised. A capability is a claim to verify like any other. Re-test the edges periodically against your own work, rather than trusting either your old map or the new brochure. When an edge moves, revise the shared page to match, so the change reaches the whole team and not just the cook who found it.

The gadget drawer holds both failures: chase every tool, and novelty eats your attention and skill never forms; refuse to update, and your map of the edges goes stale as other kitchens keep theirs current. Threading between them means staying stable on the fundamentals and current on the specifics, without becoming credulous about either.

A serious kitchen adopts the new tool that improves the food and ignores the ten that don't, because the house judges every gadget against an unchanging understanding of cooking.

# THE HABITS THAT OUTLAST THE TOOLS

## TURNING THE LIST INTO A PRACTICE

Every good cook has a way of working that is hers: the setup, the order of operations, the small fixed habits that let her own kitchen run without conscious thought. This book has handed you a set of habits: clear tickets, a stocked counter, the draft mindset, tasting before serving. A list of habits read once protects no one. The last step in using these tools well is turning that list into a practice you live by, a handful of defaults so automatic that you no longer decide them, you just work.

The move from knowing to doing is the hard part, because every failure you have been warned about happens in the gap between them. You know to verify the figure, and the deadline is close, and you skip it. You know not to paste the client's document, and it would be so much easier, and you do. The habits only protect you when they have become reflexes that hold under pressure. A habit you have to remember to follow is one you'll forget at the worst moment. A practice is what you do without deciding to.

So build it small and concrete. Pick a few defaults that fit your work and commit to them until they are automatic: a verification rule you never skip for anything that matters, a hard line on what you will never feed the tool, a standing habit of treating the first output as a draft, a short list of tasks you will not hand it. A few defaults you keep beat an elaborate system you abandon in a week, the way a cook's reliable prep beats a beautiful plan pinned to the wall and ignored.

The aim of the practice is a particular kind of ease. When the habits are in place, you stop thinking about using AI as a separate, effortful thing and just work well, with the tool as an ordinary part of your kitchen, used where it helps, checked as a matter of course, kept out where it does not belong. Fluency with these tools looks like that: not a head full of clever prompts, but a steady way of working that captures the speed without inheriting the risks. The safeguards sit inside how you cook rather than bolted on when you remember.

That quiet gap between understanding and habit is where people get hurt. Most of them aren't ignorant; they know the output can be wrong. What they never did was build the knowing into a practice, so they verify when it's convenient and skip when they're busy, and the skipped checks land where the stakes and the pressure are highest.

A cook does not become good by reading about cooking; she becomes good by cooking until the right moves are automatic and the kitchen runs itself under her hands. Which brings us, finally, to what all of this has been building toward: the kind of cook you are becoming.

## THE COOK YOU ARE BECOMING

A cook finishes training not because she has memorized every recipe, but because she has built the judgment to handle any dish, including ones she has never seen. That picture describes where this book has tried to take you. Not a collection of tricks for a particular tool, which would be stale within the year, but a way of working steady enough to handle whatever these tools become. If you have the stance, you can cook in a kitchen that has not been built yet, and that, more than any specific technique, is using AI well.

Pull the thesis of the series together. These tools are powerful, tireless, useful, and fundamentally fallible, confident whether right or wrong, with no inner sense of truth and no stake in the outcome. Using them well is therefore not a matter of finding the magic words. It takes a stance. You direct the work clearly, you verify what comes back, and you own the plate that goes out under your name. Around that, you learn where the tool is strong and where it burns, and you keep your own skills sharp enough to stay in command. The prep cook does the volume; you stay the chef. Every chapter in this book has been one face of that single stance.

The fundamentals earn their place here for the same reason. The tools will keep changing, and a book of current tricks would betray you the moment they shifted. But how these tools behave, how they fail, and how to judge what is worth handing over, these move slowly and carry across every tool you'll ever touch. You aren't leaving this book ready to use one product. You are leaving it ready to evaluate the next ten by the same questions.

Be honest about what comes next, because the stance is never finished. The temptation to over-trust the confident output, and the opposite temptation to refuse a useful tool out of fear, will both keep returning, and you will keep meeting them with the same judgment rather than retiring it. And the stakes are rising: as these tools grow more capable and we hand them more to do on their own, the cost of using them without judgment grows with them, which is the subject *AI for Dinner* takes up, where the kitchen acts instead of advising. The discipline you have built here is what that more powerful world will demand still more of.

No grand burn ends this book, only the steady one it has returned to throughout: the danger is never the tool's power and always the moment you stop bringing your own judgment. The cook who gets hurt is not the one who used the machine. It is the one who let the machine's confidence stand in for her own tasting. Keep tasting. Stay the cook who is impressed by nothing she has not checked, and you will get the speed and the help without ever handing over the thing that made the work yours.

So here is your kitchen, and it's yours. The tool is a remarkable prep cook, fast and willing and never tired, and it cannot taste, cannot care, and cannot answer for the result. You can do all three. Direct it well, check its work, keep your knives sharp, and cook. That's the whole of it, and it is enough.

# THE KITCHEN CARD, TO GO

*The whole book on one index card, the kind a cook keeps smudged and within reach. Not a system to admire and abandon, just the few habits that hold under pressure. Read straight down before a task, or find the one line you needed.*

**The stance**

1. **Stay the chef.** The tool is a fast, tireless prep cook. It does the chopping, the drafts, the volume; you direct it, judge it, decide what is good enough, and own the plate that goes out under your name.
2. **Two ways to get it wrong.** Hand over the judgment and you serve confident mistakes and lose your edge; refuse the help and you lose hours on the safe, high-volume work it does well. Run the relationship deliberately, all day.

**Setting up**

3. **Stock a small kit.** A capable general assistant, maybe a specialized tool or two, chosen by trying them on your own real tasks, not by the famous name or the latest release.
4. **Fit the pan to the cooking.** Put the tool into the work you already have, at the spots where it helps, rather than rebuilding your work around it. Start with one or two recurring tasks and earn the next from what works.
5. **Decide your data line in advance.** Whatever you paste goes to whoever runs the tool and may be kept. Keep confidential, client, personal, and regulated material out of any tool you have not cleared, and hold that as a standing rule, not a judgment made in a rush. At home nobody sets the line for you, so set it yourself, and hold it hardest for other people's details and your children's.

**The core habits**

6. **Write a clear ticket.** Brief the tool the way you would a competent colleague: the task, the audience, the form and length, the constraints, and one example of what good looks like.
7. **Stock the counter.** Most disappointing output comes from a bare prompt, not a bad one. Hand the tool the real document, the real data, the real message, the relevant background, current and correct, curated, not dumped.
8. **The first plate is a draft.** Expect raw material, not the finished answer, and steer it through a few quick rounds. Then stop and finish it in your own hand.
9. **Always taste before you serve.** Never send, publish, act on, or decide from output you have not checked. The check has four moves: circle the specifics, trace each to its source, confirm the cited thing exists, ask what is missing. The faults that cost you never show on the plate.
10. **Scale the check to the stakes.** A glance for a brainstorm; real verification for a number, a fact, or anything headed to a client or out under your name.

**The everyday dishes**

11. **Words.** A strong drafting and editing partner. Keep the voice and the facts yours; it writes a flattened average and will quietly insert claims you never made.
12. **Reduction.** Summary and synthesis save hours but drop things by the tool's sense of importance, not yours. For anything you will act on, let the summary point you back to the source, never replace it.
13. **Research and numbers.** A sound guide for the general and established; a confident inventor of the specific, the fact, the figure, the citation. Let it reason and set up the analysis, give the actual computation to a real calculator, and verify every number you serve.
14. **Writing to people.** It cannot see the relationship. Lean on it for the routine note; take the pen back for anything where a person's feelings, trust, or your standing is the real cargo.
15. **Thinking out loud.** The lowest-risk use there is: ask for fifteen options, ask it to argue against you, talk a tangle through. Then do the judging yourself, because its ideas cluster around the obvious and its first framing becomes yours before you notice.
16. **Learning.** A patient tutor, and the one place a wrong lesson goes uncaught, because not knowing is why you came. Read the explanation, then close it and do the thing badly yourself; recognition is not skill.
17. **Code and images.** Genuine help, and the same confident-wrong problem in another medium. Run and test the code; treat the image tells as a bonus already expiring, not a defense, and do not assume a generated picture is yours to use.

**What you are shown**

18. **Seeing is no longer proof.** Treat a recording as a claim, not as evidence. Confirm an urgent request through a separate, known channel, keep a family code word, and judge striking media by where it came from, not how real it looks.

**Cooking for others**

19. **What you hand off, you vouch for.** The moment your output feeds someone else's work, taste harder, and flag anything you have not verified. On a team, write the standards down, especially what never goes in and what always gets tasted, because one person's guess is the whole group's exposure.
20. **Verify first, disclose when it matters.** Disclosure is owed when it changes what the recipient thinks they are getting; verification is always the duty. Never let the one stand in for the other.

**Working with agents**

21. **Autonomy is not a switch but a dial.** More autonomy is more power and more risk together. Start any new agent hands-on, let it earn trust, and turn the dial toward hands-off only as far as proven reliability and acceptable stakes allow.
22. **Scope the job.** An agent does well with a clear goal, the right tools, a bounded

space, mostly reversible actions, and a checkable result. Vague, very long, or high-stakes jobs are where it fails.

23. **Count the whole bill.** The same arithmetic at every size: the new fit that has to beat your old way, the per-task overhead that can make doing it yourself faster, and the running cost of an agent, oversight included. The total value must beat the total cost, honestly counted.

**Staying sharp**

24. **Keep your knives sharp.** Skills fade with disuse, and judging the tool's work takes the very skill it performs. Do enough of the core work yourself to stay able to supervise.
25. **Don't trust the confident voice.** Polish is not proof, and the better the tool gets day to day, the more it lulls you. The smoothest, surest answer in a high-stakes spot earns a second look, not a pass.
26. **Know which dishes not to attempt.** Using the tool well includes the judgment not to use it: when you cannot verify and the stakes are high, when the data cannot safely go in, when the thinking is yours to do, when doing it yourself is simply faster, when the question is licensed ground, your health, your legal position, your money, the ground where the tool starts the thinking and a professional finishes it, or when the whole value of the work is that a person did it.
27. **Learn your own edges.** The general map is a starting point; draw the sharp one from your own hits and misses, trust the tool where it has earned it, and redraw the map as the tools change.
28. **Build the practice.** A few defaults made reflexive beat an elaborate system abandoned by Thursday. Pick the one or two habits that protect you most, make them automatic, and add from there.

*The danger is never the tool's power. It is the moment you stop bringing your own judgment to it. Keep tasting.*

# ALSO AT THE TABLE: THE AI AT THE TABLE SERIES

This was breakfast: the working habits, served short enough to use. The series sets more places at the same table, in the same plain-English, no-hype spirit, each book standalone and each built on the same kitchen.

*AI for Lunch* is the why behind every habit here: what these systems are, how they learned, why they are confidently wrong, and how much to trust them. If this book told you what to do, that one makes you the person who knows the reason.

*AI for Dinner* is the deep course: the turning points that built the kitchen, and then the kitchen that acts, the agents, their failures, and the controls that keep them yours. After the working kit in these pages, you are ready for the full meal.

Same kitchen, same table, more courses.

# ABOUT THE AUTHOR

Theo Kernel is an Australian technologist who has spent two decades in the IT industry, long enough to watch several "revolutions" arrive, overpromise, and settle into ordinary tools people actually use. He came to artificial intelligence the way most people will: needing to get real work done with it, and wanting to understand what he was trusting before he trusted it.

Away from the screen he is happiest in a kitchen. That is where this series' one stubborn analogy comes from, and where its central conviction was formed: that anything worth understanding can be explained at the table, in plain language, without jargon and without hype.

*AI at the Table* is his first series.

www.ingramcontent.com/pod-product-compliance
Lightning Source LLC
LaVergne TN
LVHW090330210926
848480LV00004B/1369